ARTS EDUCATION AND THE INNOVATION ECONOMY

Ensuring America's Success in the 21st Century

John M. Eger
Van Deerlin Professor of Communications
Director, Creative Economy Initiative
School of Journalism and Media Studies
San Diego State University
San Diego, California

All proceeds from the sale of this white paper will benefit arts and education in the San Diego region

" The Arts can no longer be treated as a frill. Arts education is essential to stimulating the creativity and innovation that will prove critical for young Americans competing in a global economy"

Arne Duncan

U.S. Secretary of Education
April 9, 2010
Washington, D.C.

INTRODUCTION

Fourteen years ago San Diego State University, working with the California Department of Transportation developed the concept of "smart communities"-- communities using new wired and wireless information infrastructures to connect every home, office, school, and hospital, organization and institution, large and small, to one another and through the Worldwide Web, to millions of other individuals and institutions around the world.

These infrastructures are important. But we quickly discovered smart let alone creative communities cannot exist without smart and creative people. Unfortunately too little has been clearly articulated about what makes us smart and, importantly, creative.

According to *Business Week Magazine*: "The game is changing ... it isn't just about math and science anymore (although those are surely important disciplines) it's about creativity, imagination, and, above all, innovation." [1]

If creativity and innovation will be the hallmarks of the most successful communities in the 21st century we need to know the answers to the fundamental questions of what makes us creative, innovative, and imaginative.

The effort to create a 21st century community is not so much about technology as it is about jobs, dollars and quality of life. It is about organizing one's community to reinvent itself for the new, knowledge economy and society; preparing its citizens to take ownership of their community; and, most importantly, about educating the next generation of leaders and workers to meet the global, social, political and economic challenges we face.

This commentary focuses on education and the vital role of the arts in preparing our young people for a new and uncertain future.

Although many people still believe that the arts "are nice but not necessary," it is becoming increasingly apparent that the arts are not a frill or an ancillary enrichment activity for elites. Indeed, they may be the most important aspect of a 21st century education. Our schools need the arts and an

[1]

art-infused curriculum to ensure our children's' and our country's competitiveness in the new global innovation economy.

AMERICA AT THE CROSSROADS

The challenge America faces in the wake of global competition is daunting.

We have lost our dominance in manufacturing, as well as in the provision of services like banking, accounting and insurance. Computers and Internet access can be found almost everywhere in the world, and most countries can provide such services anywhere, anytime and usually at a fraction of what it costs in the U. S.

Globalization 3.0, as author and *New York Times* columnist, Thomas Friedman calls it, is here.[2] Outsourcing jobs and off--shoring companies are commonplace. We are currently suffering what economists are euphemistically calling a "jobless recovery," and our communities and schools are facing challenges not well understood by politicians, policy makers or parents.

Twenty years ago it was fashionable to blame foreign competition and cheap labor markets abroad for the loss of manufacturing jobs in the U. S., but the pain of the loss was softened

---

2

by the emergence of a new services industry. Now, it is the service sector jobs that are being lost, and research and development, which is the life-blood of an Innovation Economy.

In just the last few years, IBM, the world's largest computer maker, acknowledged that a number of software and chip development and engineering jobs were being moved to India and China. [3] Industry stalwarts like Microsoft, Hewlett-Packard and Dell Computer announced that they, too, were either outsourcing their software development or expanding their foreign subsidiaries in China, India, the Eastern Bloc, or Russia to do the same.

Now these same companies and many others, as a matter of economic survival, have not only continued to outsource jobs but have off-shored entire divisions of their companies.

According to *SourcingMagaizne.com*, a new web site devoted to reporting on the loss of American jobs "frequently, work is off-shored in order to reduce labor expenses. Other times, the reasons for off-shoring are strategic -- to enter new markets, to tap talent currently unavailable domestically or to overcome

[3]

regulations that prevent specific activities domestically." [4]

We don't know exactly how many jobs are lost from outsourcing or off shoring. But this shift of high tech service jobs will be a permanent feature of economic life in the 21st century. And it is clear that the pervasive worldwide spread of the Internet, digitization and the availability of white-collar skills abroad--where the labor cost alone may justify the move--mean huge cost savings for the global corporations.

Making matters worse, we are nowhere near ready to capture the high ground in the new competitive environment, given the lackluster performance of our systems of education today.

Former Assistant Secretary of Education Diane Ravitch reported on the results of the latest international assessment—the Program for International Student Assessment (PISA), first performed in 2000 and repeated every three years. PISA is a worldwide evaluation of 15-year-old school pupils' scholastic performance, coordinated by the Organization for Economic Cooperation and Development (OECD) with a view to improving educational policies and outcomes.

---

[4]

She warned, "Our students scored in the middle of the pack! We are not No. 1! Shanghai is No. 1! We are doomed unless we overtake Shanghai!" She further argued: "The lesson of PISA is this: Neither of the world's two highest-performing nations (China and Finland) do what our "reformers" want to do. How long will it take before our political leaders begin to listen to educators? How long will it take before they realize that their strategies have not worked anywhere? How long will it be before they stop inflicting their bad ideas on our schools, our students, our teachers, and American education?" [5]

The *New York Times,* also reporting on the PISA tests, interviewed US Secretary of Education Arne Duncan who said: "We have to see this as a wake-up call...I know skeptics will want to argue with the results, but we consider them to be accurate and reliable, and we have to see them as a challenge to get better." He added. "The United States came in 23rd or 24th in most subjects. We can quibble, or we can face the brutal truth that we're being out-educated". [6]

The changes most policymakers and economists are talking about center around creativity and innovation[7], because

5

6

7

knowledge, broadly defined, is our salvation. We need to lead the world in new inventions, patents, products and services.

GLOBALIZATION 3.0 AND THE 21ST CENTURY WORKPLACE

Thomas Friedman's "Globalization 3.0" recognizes the tremendous growth the Internet has had literally flattening the playing field and ushering in a form of economic competition for jobs where everyone anywhere is able to compete with everyone else.

No previous telecommunications advance – not the telephone, the television, cable or even the cell phone is having more cultural and political impact on the global media landscape than the Internet. It has completely penetrated and dominated public consciousness and rapidly secured widespread public adoption.

As nations around the world awaken to the importance of creating a robust communications infrastructure, they will slowly but surely develop their Internet strategies attracting high tech service jobs that were once located in the United States. The net result of these new Internet strategies will be fewer jobs for America as nations and individuals compete for goods and services in this new global, often virtual, marketplace.

Neither a high school diploma nor a coveted degree from one of America's great universities will ensure our young people get hired. Those diplomas and degrees are worth little unless our graduates can join the ranks of the most creative and innovative knowledge workers. Creativity and innovation will be the hallmarks of the most successful companies, nations, communities and individuals.

The current worldwide financial meltdown has gutted our job market and as the dust settles, we are facing a "jobless recovery." one where corporations do well, and the stock market appears to reflect those increased earnings, but the unemployment rate remains static. The situation is dire and our systems of education are challenged as never before as our schools try to solve 21st century problems using 12th century techniques.

One problem is the failure to acknowledge the interdisciplinary world in which everything is connected to everything else, and the structure of our educational system fails to embrace this reality.

The Chronicle of Higher Education, an academic journal covering postsecondary education in the United States, recently raised the question of whether university majors are "silos" inhibiting learning.[8] Silos are one of

8

the reasons that administrators and faculty have such a difficult time making changes that count. The silos exist in K-12 and our community colleges too.

Compounding the problem, according to the U.S. Department of Labor, is that young people will "have 10 to 14 jobs by age 38." In addition, according to former Education Secretary Richard Riley, "the top 10 jobs that will be in demand (don't yet exist) and they will be using technologies that haven't been invented. In order to solve problems we don't even know are problems yet." [9]

With the proliferation of the Internet and the computerization of news archives and libraries available on the Worldwide Web, literally thousands of references are available at the click of a mouse. In an age where we are discovering that everything is connected to everything else, what we really need to do is create the interdisciplinary curriculum that emphasizes the new economy, the role of technology and the spirit of enterprise - specifically creativity and innovation.

---

[9]

ENTER THE CREATIVE AGE

A number of think tanks argue that the elements are in place for the advance of the Creative Age,[10] a period in which free, democratic nations thrive and prosper because of their tolerance for dissent, respect for individual enterprise, freedom of expression, and recognition that innovation, not mass production of low-value goods and services, is the driving force for the new economy.

The new economy's demand for creativity has manifested itself in the emergence and growth of what author Richard Florida has termed the "Creative Class," [11]

Although Florida defines this demographic group very broadly, he does a convincing job of outlining the facts of life and work in the new knowledge economy. As he points out, "every aspect and every manifestation of creativity -- cultural, technological and economic -- is inextricably linked." By tracking certain migration patterns and trends, Richard Florida did a huge service for those struggling to redefine their communities for the new knowledge economy. However, many questions remain.

10

11

Can the community, through public art or cultural offerings, enhance the creativity of its citizens? And if the new economy so desperately demands the creative worker and leader, what should schools and universities do to prepare the next generation of creative people?

Until recently, there has been only limited evidence of the connection between the arts and art-infused education and success in the postindustrial age of information.

It is now increasingly apparent that arts and art-infused, interdisciplinary and "project-based" initiatives will be the hallmarks of the most-successful schools and universities and, in turn, the most-successful and vibrant twenty-first-century communities and regions.

Those communities placing a premium on cultural, ethnic, and artistic diversity, reinventing their knowledge factories for the creative age, and building the new information infrastructures for our age, will likely burst with creativity and entrepreneurial fervor.

These are the ingredients so essential to developing and attracting the bright and creative people to generate new patents and inventions, innovative world-class products and services, and the finance and marketing plans to support them.

We now know a lot more about the brain, what makes people creative and what we need to do reinvent the curriculum, reinvent the current system of education, and redefine our very definition of education to meet the needs of the new creative and innovative economy.

THE DECADE OF THE BRAIN

With a joint resolution of Congress and a subsequent Presidential Proclamation declaring the 90's the "Decade of the Brain", research and collaboration was widely encouraged to better understand how the brain works. The National Institutes of Health, the National Institute of Mental Health, and other Federal research agencies began comparing notes with thousands of scientists and health care professionals in universities across the country.

While neuroscientists do not usually characterize functions between one hemisphere and the other, it is a fact that the left or right hemisphere of the brain dominates certain functions. Artists or those trained in the arts are usually categorized as "right brained."

A colloquialism that acknowledges the role of the right hemisphere of the brain, according to Ian McGilchrist, neuroscientist and author of *The Master and his Emissary* is that "evidence shows that the right hemisphere pays wide-open attention to the world, seeing the whole, whereas the left hemisphere is adept at focusing on a detail. New experience, whatever its kind, is better apprehended by the right hemisphere, whereas the predictable is better dealt with by the left." [12]

According to many experts, "The left hemisphere of the human brain controls language, arguably our greatest mental attribute (while) the right hemisphere is dominant in the control of, among other things, our sense of how objects interrelate in space" [13]

Our success in a new economy demanding creativity and innovation will come from nurturing both hemispheres of the brain--the whole brain--working in tandem. Author and educator Mihaly Csíkszentmihályi calls it FLOW…a " mental state of operation in which a person in an activity is fully immersed in a feeling of energized focus, full involvement, and success in the process of the activity." [14]

12

13

14

Dr Richard Restak in his book, *Mozart's Brain*[15] uses the words "plastic" and "malleable" to describe the brain. He believes that we can be creative by acquiring the right series of "repertoires"; that we can "preselect the kind of brain (we) will have by choosing richly valued experiences." In short, he and many other neuroscientists are beginning to conclude that we all have the capacity to be creative.

This fact was made clear by Robert and Michele Root-Bernstein, who researched and wrote about the things that make each of us special in some way because we followed those paths that nurture the whole brain, particularly the right hemisphere, but importantly, those things that distinguished their experiences in both cultures of art and science. [16]

During his studies as MacArthur prize fellow in La Jolla California, Root-Bernstein completed his research of 150 eminent scientists from Pasteur to Einstein. His findings were startling to those educators lobbying for more emphasis on the sciences, for he discovered that nearly all of the great inventors and scientists were also musicians, artists, writers, or poets.

15

16

In their book *Sparks of Genius,* the Root-Bernstein's point out that Galileo was a poet and literary critic. Einstein was a passionate student of the violin. And Samuel Morse, the inventor of the telegraph and father of telecommunications, was a portrait painter.

As we enter the age of the new brain, new technologies like genetic mapping and imaging are revealing to us for the first time the mysterious secrets hidden within our skulls.[17] The whole field of neuroscience has grown tremendously in the last few years, with continued research in the field of the arts and their role in education a recent area of emphasis. Importantly, we have learned that when both hemispheres of the brain are working together in harmony, we are more imaginative, creative and thus, productive.

Five years ago the Conference Board, an international non-profit business research organization, released "Ready to Work", a study which clearly agrees that "U.S. employers rate creativity and innovation among the top five skills that will increase in importance over the next five years, and rank it among the top challenges facing CEOs." [18]

---

[17]

[18]

Confirming their assessment, IBM, reported recently, "creativity is now the most important leadership quality for success in business, outweighing even integrity and global thinking." [19]

Most analysts studying the new global economy agree that the growing "creative and innovative" economy represents America's path to a brighter economic future. Whether we can all be a Picasso or Einstein is another matter. Importantly, by focusing on a curriculum that gives young people the new thinking skills they need, we can help ensue our nation's and our children's success in the new economy.

19

"America is not going to succeed through cheap labor or cheap materials, nor even the free flow of capital or a streamlined industrial base...to compete successfully, this country needs creativity, ingenuity, and innovation.

Dana Gioia, former Chairman
National Endowment for the Arts
December12, 2006
Washington, D.C.

ART AND ART INFUSION

As Gioia, has said, "to compete successfully, this country needs creativity, ingenuity, innovation." This process of transformation starts by increasing school attendance and fully engaging the student.

More than ten years ago in New York's South Bronx, the poorest congressional district in the nation and a place where only one in four children once graduated from high school, a small school called St. Augustine boasted that 95 percent of its students read at or above grade level and 95 percent met New York state academic standards.

A PBS special documentary called "Something Within Me." reported that St. Augustine made highly significant achievements despite a student population that was 100 percent minority, with many of the children living in single-parent homes in communities plagued by AIDS, crime, substance abuse and violence.

What was the secret of the school's success?

St. Augustine infused every discipline – English, math, history, science, and biology -

with dance, music, creative writing and visual arts. All the students not only excelled; they were fully engaged and reportedly, full of joy and wonder. St. Augustine, not only infused the arts into the curriculum however, they made the connections between what students were learning and the world they lived in. Sadly, as the parish was located in an extremely poor neighborhood, the school was eventually closed for financial reasons.

THE IMPORTANCE OF AN INTERDISCIPLINARY CURICULUM

In 2002, a unique consortium of arts organizations embraced the principles of a study called "Authentic Connections: Interdisciplinary Work in the Arts" to enable "students to identify and apply authentic connections, promote learning by providing students with opportunities between disciplines and/or to understand, solve problems and make meaningful connections within the arts across disciplines on essential concepts that transcend individual disciplines." [20]

Everything is connected to everything else. The interdisciplinary curriculum "encourages students to generate new insights and to synthesize new relationships between ideas." While not a manifesto for arts infusion, these

20

recommendations go far in fostering curriculum integration and offering a way for teachers of traditional, disparate disciplines to collaborate.

The Chicago-based effort: "Renaissance in the Classroom," also known as CAPE (Chicago Arts Partnership in Education), is one such model of interdisciplinary collaboration often referenced by Secretary of Education Duncan. At CAPE "teachers, artists, school administrators and parents work collaboratively to develop and share innovative approaches to teaching and learning in and through the arts." [21]

Such a multidisciplinary approach encourages leaders of young learners to see the connections between knowledge in one area and another, between a unit in mathematics and a unit in social studies, or between a unit in science and a unit in language arts. According to CAPE, "This process shows students that such thinking is possible and actually done in the real world."

FAST FORWARD TO 2011

After a decade of studying the human brain, according to the Dana Foundation we know the arts enhance math and science comprehension. We know that where art-

21

infused education is used to redesign the curriculum, one that is truly integrated, collaborative and interactive, students' attendance dramatically improves, as does performance.[22]

The National Assembly of State Arts Agencies, added to the discussion in a policy paper directed at the Obama Administration and noted, "research documents widespread acceptance of the importance of imaginative skills to innovation, and of the link between arts learning and the development of these skills."

As John Cimino, the New York State Alliance for Arts Education keynote speaker at a Conference on Creativity and Other Boundless Resources for Recession-Era Education, said; "we will achieve very little to advance our standing as a society unless each of us individually and via our institutions collectively invest wholeheartedly and with unfailing commitment in the development of the one resource native to all of us that we cannot deplete by overuse, but which in fact multiplies with use, and that is our capacity for creativity and original thinking."

22

The Massachusetts Advocates for the Arts, Sciences, and Humanities (MAASH), a statewide non-profit organization that advocates on behalf of the Massachusetts cultural community, noted in their commentary supporting the legislature's call for a "creativity index" for all the state public schools, "We have moved into an economy driven by ideas and innovation. But, are we giving our students the opportunity to develop creativity-the ability to generate ideas and then to critically evaluate potential?"

Maybe we really need to eliminate all the existing "silos" in education and infuse the curriculum again with the arts. As the Arts Education Partnership has reported, the term arts integration has evolved over the past 15 years as school districts, state arts councils, and arts organizations have experimented with various models of implementation."

However, as the Partnership noted, "Some programs and schools have chosen not to use the term at all, although descriptions of the curriculum appear to belong in this domain. Much work in the arts professional journals that could be termed integrative is labeled interdisciplinary, perhaps because, as noted in this review, the term evokes less controversy and challenge from within the arts professions." [23]

23

This is the sad truth.

The sciences and math have formulas, they have equations. As Richard Deasy, former director of the Arts Education Partnership, once complained: "the fundamental problem we confront in making the arts an unquestioned part of the learning required of students and teachers is the position of the arts in the broader culture."

Deasy suggested what's most valued in America is "muscularity" or toughness. The math and science curricula carry with them this sense of muscularity through their inherent formulas, truisms and theories. By comparison, the arts experience seems less tough, softer, and more anecdotal. [24]

BUSINESS TAKES NOTICE

Harvey White, co-founder of both Qualcomm Inc. and Leap Wireless International Inc., in a recent opinion editorial for the *San Diego Union Tribune* on the current state of education wrote, "This is not an issue about including arts because it is "nice" to do so, but rather it is an imperative because our economic future is at stake". [25]

24

25

Some businesses, which should see the threat to our economic future, understand the importance of the arts, but so far, have been only "modestly" arguing for arts integration according to The Manufacturing Institute's National Center for the American Workforce.

That could change soon, as they stated: "we believe we've only scratched the surface when it comes to the integrated arts education and STEM discussion because promising examples of integrated arts education advocacy and action are available. And assimilating STEM and art education (or art education and another business-driven education initiative) may very well be the tipping point for greater support by the business community in integrated arts education." [26]

Clearly, federal legislation indicates we are going in the wrong direction. More than three years ago, for example, then-president George W. Bush signed into law a bill called the America Competes Act, also known as the STEM initiative for Science Technology Engineering and Math.

The administration bill authorized $151 million to help students earn a bachelor's degree, help math and science teachers to get teaching credentials, and to provide additional

26

money to help align kindergarten through grade 12 math and science curricula to better prepare students for college. [27]

President Obama has called for yet a new effort called "Race to the Top," but has also called for a renewed STEM focus. As a consequence, centers and institutes for STEM are popping up across the nation. Not surprisingly, STEM is on everybody's lips and dire futures are predicted unless we all get behind STEM.

In a commentary in *The Wall Street Journal*, Chester E. Finn Jr. and Diane Ravitch, both assistant secretaries of education in the first Bush administration, complained loudly: "This is a mistake that will ill serve our children while misconstruing the true nature of American competitiveness and the challenges we face in the 21st century." [28]

In truth, we need a huge infusion of capital and a change in attitude about art and music, math and science. We need to define a well-rounded education and to make the case for its importance in a global innovation economy. As demand for a new workforce to meet the challenges of a global knowledge economy is rapidly increasing, few things could be as

27

28

important in this period of our nation's history as art and art-infused education.

As Qualcomm's White, who was responsible for hiring thousands of engineers, has observed, engineers need the arts as well as sciences; or American's will not be as creative and innovative as America need to be in the global economy. "The arts are an integral and necessary part of educating our future innovators so (our workers) can compete successfully in the forthcoming world economy...we need to get business, government and media to connect the dots between arts education and economic success".

At the Federal level, STEM needs to recognize the vital role of the arts. The Elementary and Secondary Education Act (ESEA), sometimes called the No Child Left Behind (NCLB) act, also needs to be changed to take into account the importance of creativity, provide the teacher training necessary to change the curriculum, and to infuse the curriculum with the arts,

States and counties cannot wait for the Federal government to act however. Together they must redefine the purpose of public education, reinvent our current systems of education, and put creativity at the top of the agenda for action.

CONCLUSION

During the Clinton Administration democratic strategist James Carville was fond of saying, "It's the economy, stupid."

Much the same could be said today.

Stimulus funds and all the federal policies in the world will not help if all we do is prop up the old economy. It is rather the new economy, the creative economy that is begging for attention.

We know that creativity and innovation are the hallmarks of the most successful economies. They represent America's future. We know, too, that educating our young people to have the new thinking skills to become productive members of the new creative and innovative workforce is vital.

However, our current system of education is not up to this challenge and produces students who perform well below their peers in most nations of the world according to the recent results of The Program for International Student Assessment (PISA).

As U.S. Secretary of Education Arne Duncan further acknowledged responding to the PISA results, "We have to see this as a wake-up call...I know skeptics will want to argue with the results, ' he said 'but we consider them to be accurate and reliable, and we have to see them as a challenge to get better...The United States came in 23rd or 24th in most subjects. We can quibble, or we can face the brutal truth that we're being out-educated".

An art infused integrated curriculum, is a logical answer to nurturing both hemispheres of the brain and producing the kinds of skills we desperately need to compete. While there are many things we must do to reinvent our schools for the new economy, we must take afresh look at the critical role of the arts in transforming our curriculum.

As a whole new economy based upon creativity and innovation emerges, the importance of reinventing business strategies, corporations, communities, and importantly, our schools, is critical.

We need to redesign our K-12 and college curricula to focus on preparing students for this new competition if we are to survive, let alone succeed, in this new global economy.

REFERENCES

1) "Get Creative: How to build Innovative Companies," *BusinessWeek*, August 1, 2005. http://www.businessweek.com/magazine/content/05_31/b3945401.htm

2) Friedman, Thomas L., The World Is Flat: A Brief History of the Twenty-first Century, New York: Farrar, Straus and Giroux, 2005.

3) "IBM Cuts Jobs as It Seeks Stimulus Money", Bloomberg BusinessWeek, March 25, 2009.
http://www.businessweek.com/bwdaily/dnfl ash/content/mar2009/db20090325_626883. htm

4) "What is Outsourcing", SOURCINGmag.com, January 22, 2011.
http://www.sourcingmag.com/

5) "The Real Lessons of PISA", Education Week, December 14, 2010.
http://blogs.edweek.org/edweek/Bridging-Differences/2010/12/the_real_lessons_of_pi sa.html

6) "Top Test Scores from Shanghai Stun Educators", The New York Times, December 7, 2010.
http://www.nytimes.com/2010/12/07/educat ion/07education.html

7) "Ready to Innovate", The Conference Board, October 2010.
www.artsusa.org/pdf/information.../policy.../r eady_to_innovate.pdf

8) "How to Create a Problem-Solving Institution", The Chronicle of Higher

Education, August 20, 2010.
http://chronicle.com/article/How-to-Create-a/124153/

9) Robert Jones, Kathryn Scanland, Steve Gunderson, The Jobs Revolution: Changing How America Works, Copywriters Inc., a division of the Greystone Group, Inc., 2005.

10) Gene D. Cohen, The Creative Age: Awakening Human Potential in the Second Half of Life, Avon Books, a division of Harper Collins, 2000.

11) Iain McGilchrist, "The Battle if the Brain", *The Wall Street Journal*, January 2, 2010.
http://online.wsj.com/article/SB1000142405 2748704304504574609992107994238.html

13) Peter F. MacNeilage, Lesley Rogers, and Giorgio Vallortgara, "Evolutionary Origins of Your Right and Left Brain", *Scientific American*, July 2009.

14) Csikszentmihalyi, Mihaly, Creativity: Flow and the Psychology of Discovery and Invention, Harper Perennial, 1996.

15) Restak, Richard, M.D., The New Brain, Rodale, 2003.

16) Root-Bernstein, Robert and Michele, Sparks of Genius: The 13 Thinking Tools of

the World's Most Creative People, Houghton Mifflin Company, 1999.

17) Pink, Daniel H., A Whole New Mind: Moving from the Information Age to the Conceptual Age, Riverhead Books, 2009.

18) "Are They Really Ready To Work", The Conference Board, October 2006. http://www.conference-board.org/publications/publicationdetail.cfm?publicationid=1218

19) "The Most Important Leadership Quality for CEO's? Creativity", *FAST COMPANY*, May 18, 2010. http://www.fastcompany.com/1648943/creativity-the-most-important-leadership-quality-for-ceos-study

20) "Authentic Connections: Interdisciplinary Work in the Arts", Consortium of National Art Education Associations (AATE, MENC, NAEA, NDEO), 2002. www.eric.ed.gov/ERICWebPortal/recordDetail?accno=ED470397

21) CAPE, Chicago Arts Partnership in Education, January 2011. http://www.capeweb.org/

22) "Neuroeducation: Learning, Arts, and the Brain", THE DANA FOUNDATION, January 2011.

http://www.dana.org/news/publications/publication.aspx?id=23964

23) Burnaford, Gail, with Brown, Sally, Doherty, James, and McLaughin, James, "Arts Integration: Frameworks, Research, and Practice, Arts Education Partnership, October 29, 2007.
www.aep arts.org/files/publications/arts_integration_book_final.pdf

24) Larson, Gary O., "American Canvas,' National Endowment for the Arts, 1997. *www.nea.gov/pub/AmCan/AmericanCanvas.pdf*

25) White, Harvey, "Arts and the Innovation Gap", San Diego Union Tribune, March 11, 2010.
http://www.signonsandiego.com/news/2010/mar/11/arts-and-the-innovation-gap/

26) "America Competes Act", Conference Report, House of Representatives, 2007.
http://www.govtrack.us/congress/bill.xpd?bill=h110-2272

27) "Getting Past Either-Or", A Feasibility Study: Arts, Innovation, and the Role of Business Champions, August 2008.
sbo.nn.k12.va.us/art/ArtsInEducation082608.pdf

FURTHER READING

1) McGilchrist, Ian, The Master and his Emissary: The Divided Brain and the Making of the Western World, Yale University Press, December 15, 2009.

2) Robinson, Sir Ken, The Element: How Finding Your Passion Changes Everything, Viking, 2009.

3) Florida, Richard, The Rise of the Creative Class: and How it's Transforming Work, Leisure, Community and Everyday Life, New York: Basic Books, 2004.

4) Howkins, John, The Creative Economy: How People Make Money from Ideas, Penguin, 2001.

5) Eisner, Elliot E., The Arts and the Creation of Mind, Yale University press, 2002.

6) Shlain, Leonard, Art & Physics: Parallel Visions in Space, Time, and Light, Viking, 1977.

7) Shlain, Leonard. The Alphabet and The Goddess: The Conflict Between Word and Image, Viking, 1998.

ABOUT THE AUTHOR

John M. Eger, Lionel Van Deerlin Endowed Chair of Communications and Public Policy and Director of the Creative Economy Initiative at San Diego State University, is an author and lecturer on the subjects of creativity and innovation, education and economic development.

A former Advisor to the President and Director of the White House Office of Telecommunications Policy he helped spearhead the restructuring of America's telecom Industry and was Senior Vice President of CBS, which opened China to commercial television. More recently he served as Chair of California Governor's Commission on Information Technology; Chair of the Governors Committee on Education and Technology; and Chair of San Diego Mayor's "City of the Future" Commission. He recently authored the seminal "Guidebook for Smart Communities", a "how to" for communities struggling to compete in the age of the Internet; and "The Creative Community: Linking Art, Culture, Commerce and Community", a call to action to reinvent our communities for the Creative Age.

We are living through the closing chapters of the established and traditional way of life. We are in the early beginnings of a struggle to remake our civilization. It is not a good time for politicians. It is

a time for prophets and leaders and explorers and investors and pioneers, and for those who are willing to plant trees for their children to sit under."

Walter Lippmann

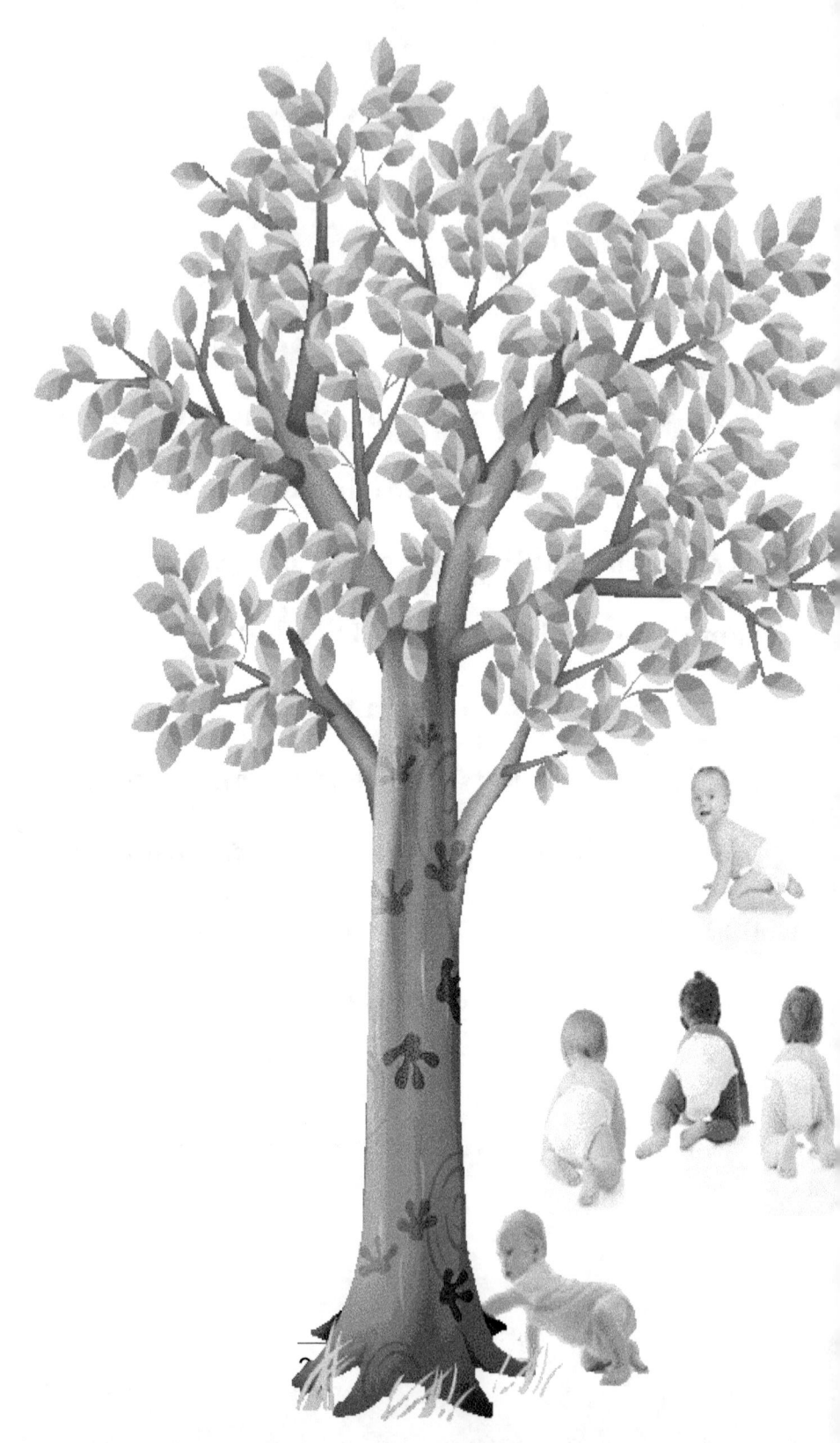